math standards workout

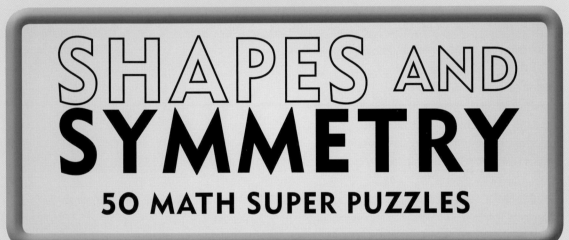

SHAPES AND SYMMETRY

50 MATH SUPER PUZZLES

By Thomas Canavan

rosen publishing's
rosen central

This edition first published in 2012 by The Rosen Publishing Group, Inc.
29 East 21st Street, New York, NY 10010

Author: Thomas Canavan
Editor: Joe Harris
Design: Jane Hawkins
Cover design: Jane Hawkins

Library of Congress Cataloging-in-Publication Data

Canavan, Thomas, 1956-
Shapes and symmetry : 50 math super puzzles / Thomas Canavan.
p. cm. — (Math standards workout)
Includes bibliographical references and index.
ISBN 978-1-4488-6676-2 (library binding : alk. paper) — ISBN 978-1-4488-6683-0 (pbk. : alk. paper) — ISBN 978-1-4488-6689-2 (6-pack : alk. paper)
1. Geometry—Juvenile literature. 2. Shapes—Juvenile literature. 3. Symmetry (Mathematics)—Juvenile literature. 4. Mathematical recreations—Juvenile literature. I. Title.
QA445.5.C35 2012
516'.15—dc23
2011028988

Printed in China
SL002077US

CPSIA Compliance Information: Batch #W12YA. For further information, contact Rosen Publishing, New York, New York, at 1-800-237-9932.

Contents

Introduction

Why do you need this book?

Whether you are a geometry genius or struggle with symmetry, there is always room for improvement. This book targets the parts of your brain responsible for shapes and symmetry and gives them a workout. These 50 puzzles will help you to build up the skills involved in finding the balance and order within shapes. They will help you to improve both in the areas where you feel most confident and in those areas where you feel weakest.

How will this book help you at school?

Shapes and Symmetry complements the National Council of Teachers of Mathematics (NCTM) framework of Math Standards, providing an engaging enhancement of the curriculum in the following areas:

> *Geometry: Analyze Characteristics and Properties of Two- and Three-Dimensional Geometric Shapes and Develop Mathematical Arguments about Geometric Relationships*
> *Geometry: Apply Transformations and Use Symmetry to Analyze Mathematical Situations*
> *Geometry : Use Visualization, Spatial Reasoning, and Geometric Modeling to Solve Problems*

Why have we chosen these puzzles?

This *Math Standards Workout* title features a range of interesting and absorbing puzzle types, challenging students to master the following skills to arrive at solutions:

- Identify, compare, and analyze attributes of two- and three-dimensional shapes and develop vocabulary to describe the attributes: e.g. Tile Twister, Hexagony
- Classify two- and three-dimensional shapes according to their properties and develop definitions of classes of shapes such as triangles and pyramids: e.g. Mini Sudoku, Pyramid Plus
- Investigate, describe, and reason about the results of subdividing, combining, and transforming shapes: e.g. Tile Twister
- Explore congruence and similarity: e.g. Hexagony, Tile Twister, Mini Sudoku

NOTE TO READERS

If you have borrowed this book from a school or classroom library, please respect other students and DO NOT write your answers in the book. Always write your answers on a separate sheet of paper.

Tile Twister

Place the eight tiles into the puzzle grid so that all adjacent numbers on each tile match up. Tiles may be rotated through 360 degrees, but none may be flipped over. Write your answers on a separate sheet of paper.

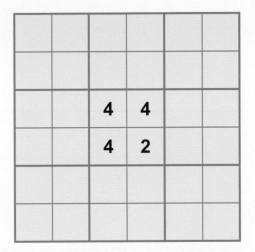

2	4		1	3		1	3		1	2
2	3		2	4		4	4		4	4

2	2		3	4		4	1		4	3
1	3		1	3		2	4		1	3

What's the Number?

In the diagram below, what number should replace the question mark? Write your answer on a separate sheet of paper.

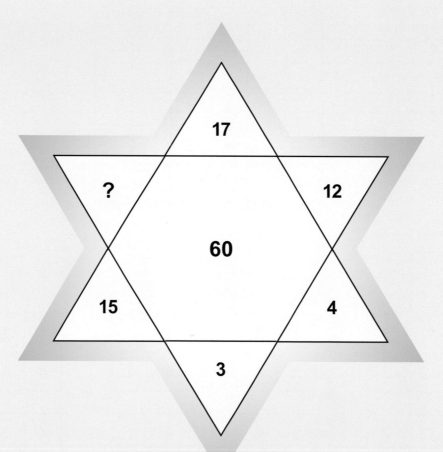

Mini Sudoku

Every row, column, and each of the four smaller boxes of four squares should contain a different number from 1 to 4 inclusive. Some numbers are already in place. Can you complete the grid? Write your answers on a separate sheet of paper.

3

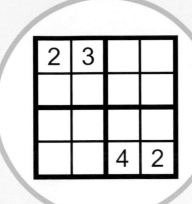

4

5

Every row, column, and each of the six smaller boxes of six squares should contain a different number from 1 to 6 inclusive. Some numbers are already in place. Can you complete the grid? Write your answers on a separate sheet of paper.

Number Path

Copy out this puzzle. Working from one square to another, horizontally or vertically (never diagonally), draw paths to pair up each set of two matching numbers. No path may be shared, and none may enter a square containing a number or part of another path. Write your answers on a separate sheet of paper.

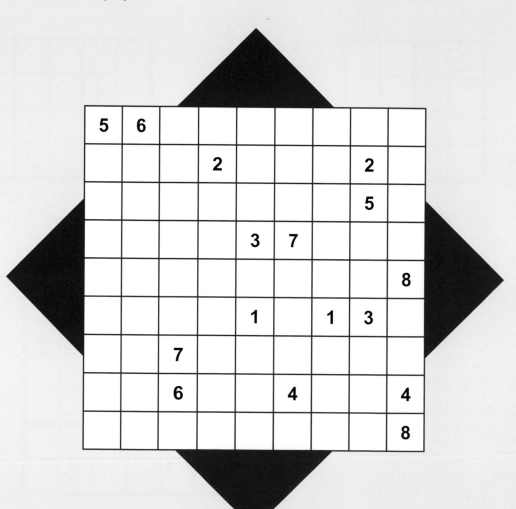

Hexagony

Can you place the hexagons into the grid, so that where any hexagon touches another along a straight line, the number in both triangles is the same? No rotation of any hexagon is allowed! Write your answers on a separate sheet of paper.

Pyramid Plus

The number in each circle is the sum of the two numbers below it. Just work out the missing numbers in every circle! Write your answers on a separate sheet of paper.

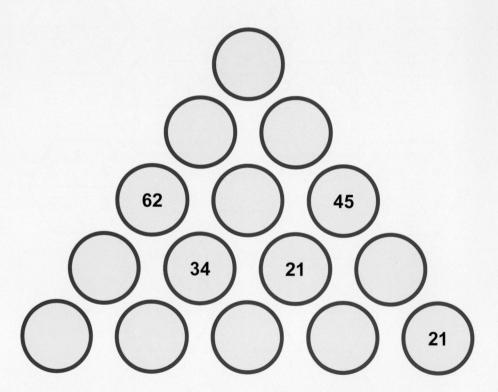

Tile Twister

Place the eight tiles into the puzzle grid so that all adjacent numbers on each tile match up. Tiles may be rotated through 360 degrees, but none may be flipped over. Write your answers on a separate sheet of paper.

9

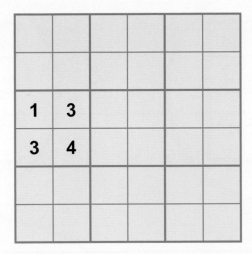

1	3				
3	4				

4	1
4	2

4	2
1	1

2	2
4	1

2	1
3	2

3	3
1	4

2	2
3	4

4	2
1	3

4	1
1	2

Mini Sudoku

Every row, column, and each of the four smaller boxes of four squares should contain a different number from 1 to 4 inclusive. Some numbers are already in place. Can you complete the grid? Write your answers on a separate sheet of paper.

10

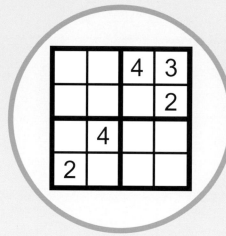

11

12

Every row, column, and each of the six smaller boxes of six squares should contain a different number from 1 to 6 inclusive. Some numbers are already in place. Can you complete the grid? Write your answers on a separate sheet of paper.

What's the Number?

In the diagram below, what number should replace the question mark? Write your answer on a separate sheet of paper.

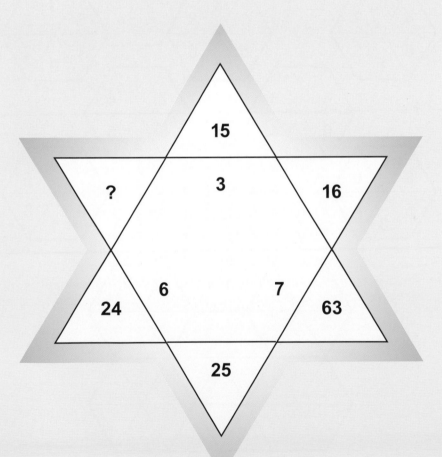

Hexagony

Can you place the hexagons into the grid, so that where any hexagon touches another along a straight line, the number in both triangles is the same? No rotation of any hexagon is allowed! Write your answers on a separate sheet of paper.

Pyramid Plus

The number in each circle is the sum of the two numbers below it. Just work out the missing numbers in every circle! Write your answers on a separate sheet of paper.

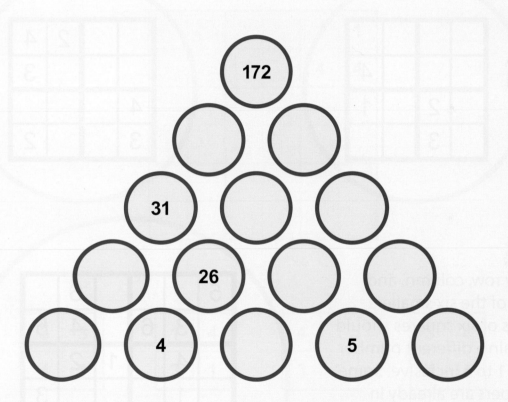

What's the Number?

In the diagram below, what number should replace the question mark? Write your answer on a separate sheet of paper.

34

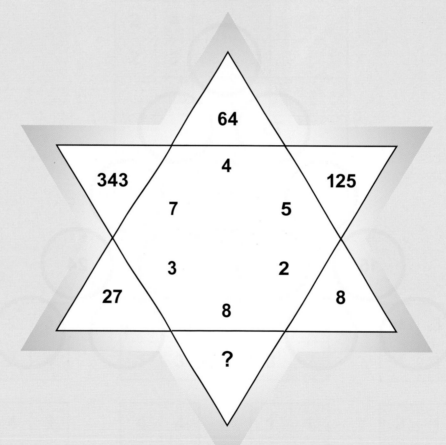

Hexagony

Can you place the hexagons into the grid, so that where any hexagon touches another along a straight line, the number in both triangles is the same? No rotation of any hexagon is allowed! Write your answers on a separate sheet of paper.

Hexagony

Can you place the hexagons into the grid, so that where any hexagon touches another along a straight line, the number in both triangles is the same? No rotation of any hexagon is allowed! Write your answers on a separate sheet of paper.

42

Mini Sudoku

Every row, column, and each of the four smaller boxes of four squares should contain a different number from 1 to 4 inclusive. Some numbers are already in place. Can you complete the grid? Write your answers on a separate sheet of paper.

43

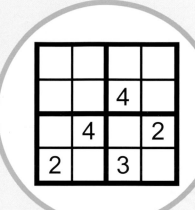

44

45

Every row, column, and each of the six smaller boxes of six squares should contain a different number from 1 to 6 inclusive. Some numbers are already in place. Can you complete the grid? Write your answers on a separate sheet of paper.

Hexagony

Can you place the hexagons into the grid, so that where any hexagon touches another along a straight line, the number in both triangles is the same? No rotation of any hexagon is allowed! Write your answers on a separate sheet of paper.

Word Puzzles

49

Clock-Watching

Draw two lines across the face of a clock to create three parts, making sure that the sums of the numbers in all three parts are the same.

50

Mirror Image

Jacob's brother was ten years old in 2011. The day he was born was a palindrome—the numbers read the same backward and forward. On what day and year was he born? You should write your date in this format: MM/DD/YYYY.

36

3	4	1	2
2	1	4	3
4	2	3	1
1	3	2	4

37

3	2	1	4
1	4	3	2
2	3	4	1
4	1	2	3

38

5	1	3	6	2	4
2	6	4	1	3	5
3	2	6	4	5	1
4	5	1	3	6	2
6	4	5	2	1	3
1	3	2	5	4	6

39

A grid puzzle with the numbers 2, 2, 3, 7, 6, 4, 5, 9, 5, 8, 1, 6, 9, 8, 4, 3, 1, 7.

40

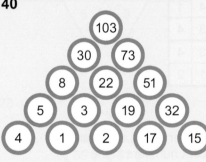

41

16 – Working clockwise from the top, the numbers represent the running total of numbers in the preceding two points of the star, ending with the central number.

42

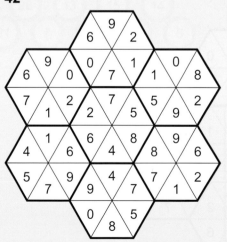

43

4	3	2	1
1	2	4	3
3	4	1	2
2	1	3	4

44

2	3	4	1
1	4	3	2
3	2	1	4
4	1	2	3

45

3	1	5	4	2	6
2	6	4	5	3	1
5	3	2	6	1	4
6	4	1	2	5	3
1	5	6	3	4	2
4	2	3	1	6	5

46

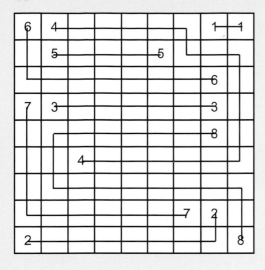

47

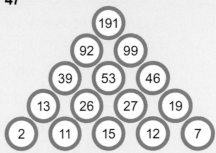

48

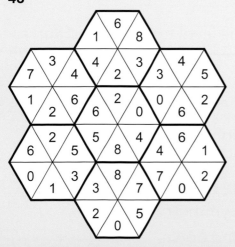

49

Draw lines diagonally, the first starting between 10 and 11 and sloping down to between 2 and 3; the second between 8 and 9 down to between 4 and 5. Each section adds up to 26.

50

October 2, 2001 (10/02/2001)

Glossary

adjacent	Close to or—more commonly—next to.
clockwise	A circular movement that goes in the same direction that a clock's hands travel.
column	A line of objects that goes straight up and down.
diagonal	Moving in a slanted direction, halfway between straight across and straight down.
diagram	A drawing or outline to explain how something works.
grid	A display of crisscrossed lines.
hexagon	A six-sided object.
horizontal	A direction that is straight across.
inclusive	Including both ends of a series (2 to five inclusive means 2, 3, 4, and 5).
matching	Exactly the same as.
mini	Small (an informal word).
pyramid	A triangular shape with one side level to the ground and a point at the top.
rotate	Travel in a circular motion.
rotation	Circular motion.
row	A line of objects that goes straight across.
shared	Having something the same as something else.
symmetry	Looking the same on either side of an imaginary line (a mirror image is an example of symmetry).
vertical	A direction that is straight up and down.

Further Information

For More Information

Consortium for Mathematics (COMAP)
175 Middlesex Turnpike, Bedford, MA 01730
(800) 772-6627 http://www.comap.com/index.html
COMAP is a nonprofit organization whose mission is to improve mathematics education for students of all ages. It works with teachers, students, and business people to create learning environments where mathematics is used to investigate and model real issues in our world.

MATHCOUNTS Foundation
1420 King Street, Alexandria, VA 22314
(703) 299-9006 https://mathcounts.org/sslpage.aspx
MATHCOUNTS is a national enrichment, club, and competition program that promotes middle school mathematics achievement. To secure America's global competitiveness, MATHCOUNTS inspires excellence, confidence, and curiosity in U.S. middle school students through fun and challenging math programs.

National Council of Teachers of Mathematics (NCTM)
906 Association Drive, Reston, VA 20191-1502
(703) 620-9840 http://www.nctm.org
The NCTM is a public voice of mathematics education supporting teachers to ensure equitable mathematics learning of the highest quality for all students through vision, leadership, professional development and research.

Web Sites

Due to the changing nature of Internet links, Rosen Publishing has developed an online list of Web sites related to the subject of this book. This site is updated regularly. Please use this link to access this list:

> http://www.rosenlinks.com/msw/shap

Further Reading

Abramson., Marcie F. *Painless Math Word Problems.* New York, NY: Barron's Educational Series, 2010.

Ball, Johnny. *Why Pi?* New York, NY: Dorling Kindersley, 2009.

Fisher, Richard W. *Mastering Essential Math Skills: 20 Minutes a Day to Success (Book Two: Middle Grades/High School).* Los Gatos, CA: Math Essentials, 2007.

Lewis, Barry. *Help Your Kids With Math: A Visual Problem Solver for Kids and Parents*. New York, NY: DK Publishing, 2010.

Overholt, James and Laurie Kincheloe. *Math Wise! Over 100 Hands-On Activities That Promote Real Math Understanding*. New York, NY : Jossey-Bass, 2010.

Index